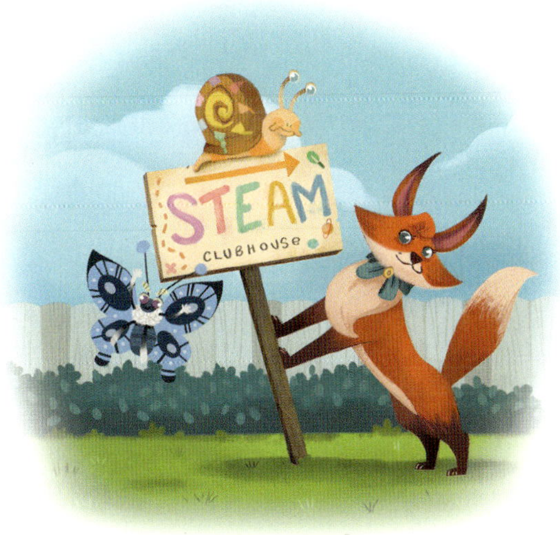
STEAM
CLUBHOUSE

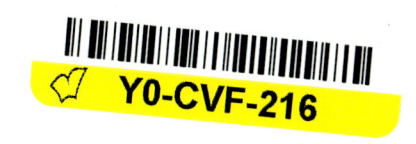

*To the little scientists reading this book:
You can do it too. You belong here.* —J.A.

To my husband, Roee, my fellow traveler around the world —D.J.K.

To my parents . . . they know what they did. —J.P.

Room to Read would like to thank Tatcha™ for their generous
support of the STEAM-Powered Careers collection.

TATCHA

Copyright 2022 Room to Read. All rights reserved.

Written by Jocelyn Argueta
Featured scientists: Dr. Dieuwertje "DJ" Kast and Jocelyn Argueta
Illustrated by Janet Pagliuca
Edited by Carol Burrell
Photo research by Kris Durán
Series art direction and design by Christy Hale
Series edited by Carol Burrell, Jamie Leigh Real, Jocelyn Argueta, and Deborah Davis
Copyedited by: Debra Deford-Minerva and Danielle Sunshine

ISBN 978-1-63845-057-3

Manufactured in Canada.

MIX
Paper | Supporting responsible forestry
FSC® C011825

10 9 8 7 6 5 4 3 2

Room to Read
465 California Street #1000
San Francisco, California 94104
roomtoread.org

Room to Read

World change starts with Educated children.©

STEAM-Powered Careers
POLAR SCIENCE

by **Jocelyn Argueta**

featured scientists: **Dr. Dieuwertje "DJ" Kast** and **Jocelyn Argueta**

illustrated by **Janet Pagliuca**

Room to Read

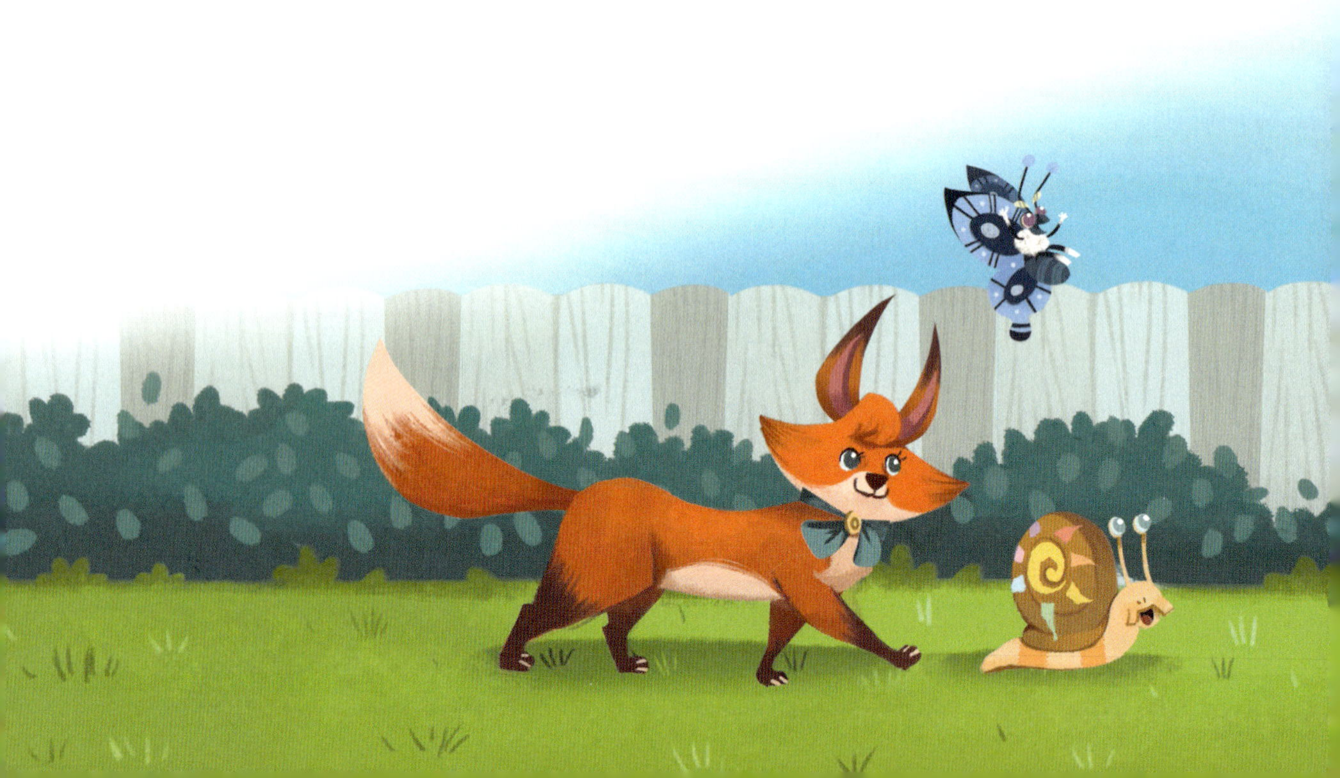

Contents

Explore Polar Science with Mia and Sunny	6
What Is Polar Science?	22
Meet the Scientists	24
Learn More about Polar Science	30
Word List	34

The day has finally arrived for Mia's big adventure. "Where are you going?" asks her snail friend Sunny. Mia points to the map. "Just to the coolest places on Earth—the North and South Poles."

Sunny's eyes widen in amazement. "Oh! I've never been there before."

Mia puts on her winter boots. "It's where polar scientists go to study things like the stars and the planet. You should come with me and see for yourself!"

Sunny can't wait to get there! Mia places him on her shoulder, and they head out the door.

Polar Science 7

Their first stop is the Arctic **tundra**. It's a large frozen plain surrounding the **North Pole**. It's buzzing with huge mosquitoes. BUZZ! BUZZ! BUZZ! Mia and Sunny scramble to put on their bug gear before they get bitten.

"Phew!" Sunny exclaims. "That was a close one."

Mia feels like she's stepping on top of basketballs. The ground is covered in grass clumps called **tussocks**, which makes it hard to walk around.

"Let's take the helicopter to the research site," she suggests. "It will get us there faster."

From the sky, Mia and Sunny take in the gorgeous view and wave to all the musk oxen in the field.

A lot of the Arctic is covered in **permafrost**, which is soil that has been frozen for two years or more.

Polar Science 9

They land next to a lake. Mia starts rummaging through her bag. "Oh, I get it. We're going swimming!" Sunny cheers.

Mia giggles. "No! We're collecting water samples with different **microbes** so we can go back to the lab and learn more about them."

"Microbes? Sounds like new friends! I would love to introduce myself," says Sunny.

Mia puts on her gloves and starts collecting water from the lake. "OK, but you'll need to use a **microscope**. The microbes are so small, we can't see them without one."

Sunny looks puzzled. "Why would polar scientists travel all the way to the Arctic just to collect microbes from the water?"

The microbes are collected in small glass containers. Each container can hold thousands of microbes.

Polar Science 11

Active layer

Permafrost with low ice content

Ice wedge

Mud deposited by thawing

Ice-rich permafrost (ground ice)

Mia points out the permafrost. "Actually, these microbes used to be frozen. Since the planet is getting warmer, the ice has melted, and now they're in the water. Polar scientists come here to collect them. Microbes help us understand how fast the Arctic ice is melting, because that can affect the whole world—now and in the future."

STEAM-Powered Careers

Sunny is surprised. "I never knew something so small could make such a big difference!"

Mia packs up her samples and equipment. They take one last look at the Arctic tundra and move on to their next stop.

During the summer months, the Arctic has more mosquitoes than anywhere else on the planet. They lay their eggs on or near shallow pools of water that form when the permafrost melts.

They leave behind the buzzing of mosquitoes and go to . . .

Polar Science 13

. . . the silence of the South Pole.

Sunny looks around and sees miles and miles of ice in every direction. "This place looks like a giant freezer! Where are all the animals?"

"Welcome to the South Pole," Mia says proudly, "the farthest south you can go! Scientists who visit this place live in the station, which is like a little indoor town. It's too cold for animals or plants here, but you can find plenty of them along the coast of Antarctica."

Mia starts up the snowmobile and hops on. "Come on, Sunny," she says. "I want to show you my favorite spot."

The flags outside the South Pole station represent the original twelve countries that signed the Antarctic Treaty, an agreement to protect Antarctica for science.

Polar Science 15

There are more than 5,000 sensors buried in the ice around the IceCube Laboratory. It took seven years to build it.

16 STEAM-Powered Careers

In about four minutes, they arrive at a big blue building. The ice crunches under Mia's boots as she walks over to it. "This is the **IceCube Laboratory**," she says.

Sunny imagines an ice-cold drink. "If these are the scientists who discovered ice cubes, I'm going to need their autographs."

Mia laughs. "They don't study ice cubes. They study the universe. The coolest part is that the equipment for the whole experiment is buried more than a mile under the ice. Other telescopes nearby can be used to see the stars and outer space in different ways."

Sunny is confused. "But we can see the stars at home. Why do scientists come all the way to the South Pole to do that?"

"The South Pole is a great place to study the stars," says Mia. "It's on top of a giant ice sheet that is 9,300 feet high. It's very dry here, so the skies are clear, and it's completely dark for six months of the year."

Mia and Sunny stand in the green glow from the southern lights in the sky.

"This is the most magical freezer I've ever seen," Sunny says in amazement.

At the South Pole, the southern lights, also known as the **aurora australis**, happen when particles from the sun hit gas molecules in the sky and make them glow.

Polar Science

20 STEAM-Powered Careers

Back at the clubhouse, Mia shares photos of their big polar science adventure.

Sunny can't wait to tell their friends about all the science he learned and why some amazing experiments can take place only in two special locations—the North and South Poles!

What is Polar Science?

Mia and Sunny have only given us the tip of the iceberg that is polar science. Before we put on our boots and tag along with **Dr. DJ** and **Jocelyn** to find out more about a day on opposite sides of the world, here is some information that will be helpful at the poles.

Polar science is the study of the frozen places at the top and bottom of the world. Earth has two poles: the North Pole and the South Pole. The North Pole is in the Arctic region, and the South Pole is on the continent of Antarctica.

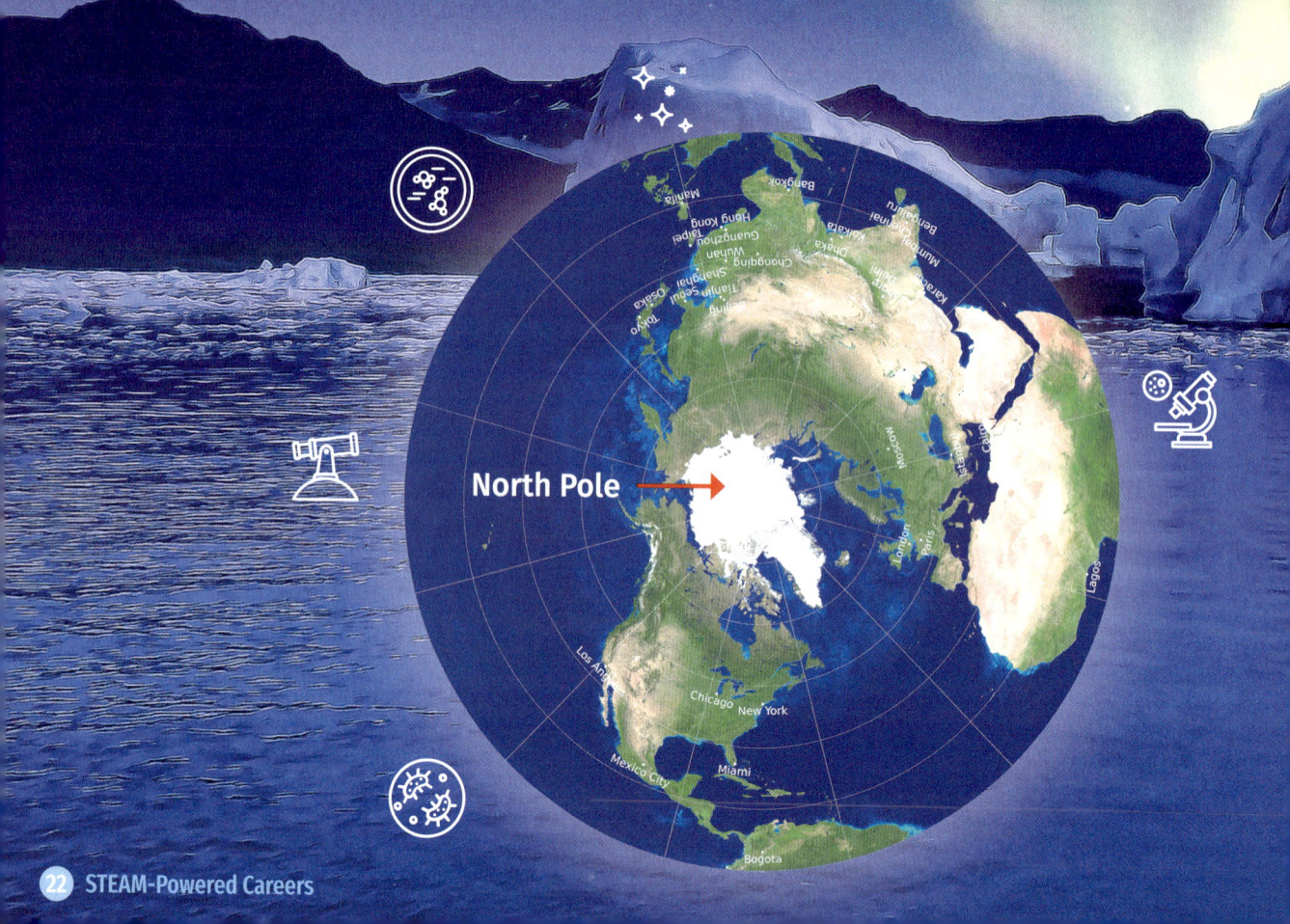

North Pole

22 STEAM-Powered Careers

Polar scientists go to the Arctic and the Antarctic to study the ice, animals, air, and water. The ice at the poles helps balance Earth's climate, which is the weather that happens over a long period of time. By studying the poles, we learn more about our entire planet!

Dr. DJ and **Jocelyn** are both polar scientists doing some fascinating work in two different extreme environments. Let's ask them some questions, and then they'll show us their "offices."

South Pole

Polar Science 23

Meet the Scientists

Dr. DJ in her polar dress at 11 p.m. at the North Pole

Jocelyn all bundled up at the South Pole

STEAM-Powered Careers

Dr. DJ

I earned a bachelor's degree in biology, a master's in marine environmental biology, and a master's in teaching, and I have a single-subject science credential. I also earned a doctorate (EdD) in teacher education in multicultural societies (TEMS).

Fun fact: I have a big collection of science dresses.

Jocelyn

I earned a bachelor's degree in biochemistry and molecular biology.

Fun fact: I love penguins and live theater.

> What is your favorite thing about the Arctic?

> I love learning new things and seeing new places.

> What is your least favorite thing about the Arctic?

> The mosquitoes and their constant buzzing!

> What is your favorite thing about the Antarctic?

> It feels like an alien planet!

> What is your least favorite thing about the Antarctic?

> Having to wear twenty pounds of clothes to stay warm and safe outside at the South Pole.

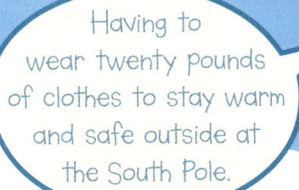

Polar Science

A Day at the North Pole

After spending a night on the tundra, we wake up in a green Arctic tent. My fellow scientists and I put on all our bug gear to protect us from the huge mosquitoes.

We take a quick fifteen-minute ride into the field and enjoy the gorgeous view. The colors are bright and the air is crisp.

STEAM-Powered Careers

We collect water samples from the lakes and rivers and take them to the lab.

Back at Toolik Field Station, the water samples are tested to find out which microbes were collected and from where.

At the end of the day, we unwind by taking photos of the landscape or visiting an aufeis, a giant piece of layered ice. Just watch out for the mosquitoes!

Polar Science 27

A Day at the South Pole

During the summer months, it's always sunny and bright because the sun is up 24 hours a day.

Breakfast is served in the galley with a full view of the twelve flags outside the station. The flags form the **Ceremonial South Pole**. This is also a great time to see how overcast and cold the day might be.

Before stepping outside, we need to put on all our extreme-cold weather gear. This includes a warm shirt and pants, a coverall, boots, gloves, and a red parka.

To get to the laboratory, we can walk or drive on our snowmobile. Because of the cold, the snowmobile might take a while to start but we're getting better at it with practice!

At the IceCube Laboratory, scientists receive information from more than 5,000 round instruments buried below the ice. The information they collect tells us about small particles called neutrinos coming from space. Some neutrinos come from exploding stars!

At the end of the day, we can walk to the **Geographic South Pole**, which is physically the southernmost point on the globe.

Most scientists and staff visit the South Pole during the summer months. Only a few people, called winterovers, stay through the winter months—when every day is 24 hours of darkness!

Polar Science 29

Polar science can feel like a big group project

where everyone is an important part of the team!

STEAM Careers in Polar Science

The cool thing about polar science is that it brings together people with different talents.

To learn more about the land or animals, you can study geology, environmental science, or biology. These experts will tell you that the Arctic is a cold, grassy tundra, while the South Pole is so dry, it's considered a frozen desert.

At the South Pole, scientists who study physics turn their eyes to the sky ... and beyond. They use telescopes or sensors like the ones under the IceCube Laboratory to learn more about outer space.

In both the Arctic and the Antarctic, engineers help design machines and buildings that can be used in these extreme locations.

Since not many people go to the polar regions, artists and writers who do go can also help share this information with the world. They can take photos, create drawings, and write stories about the special things they see and experience.

Polar Science

The Future of Polar Science

Even though the polar areas are far away from where most people live, they still affect the rest of the world in a surprising number of ways.

Polar science at both the North Pole and the South Pole will help us protect the planet and plan for the future. Polar science is part of some of the most important work of all: understanding climate change and the damage to our planet's environment.

Polar scientists are brave and curious. They travel to distant places to do science experiments that can't be done anywhere else. One day, you could be a polar scientist too!

Do You Want to Be a Polar Scientist?

You can start your journey to the North and South Poles today. Here are some tips to get you on your way:

- **Be curious about the world around you.** Ask questions and learn about the things on Earth that make you excited.

- **Ask a parent, teacher, or guardian to explore the outdoors with you.** Polar scientists are always ready for an adventure!

- **Ask a librarian to help you find books and experiments you can try at home.** Look for topics like polar animals, climate change, and astrophysics.

Word List

aurora australis: beautiful glowing lights in the sky at the South Pole, caused by particles from the sun reaching the Earth. Also called the Southern Lights.

Ceremonial South Pole: a location with twelve flags to remember the original twelve countries that signed the Antarctic Treaty, an agreement to protect Antarctica

Geographic South Pole: the location that is exactly the southernmost point on Earth

IceCube Laboratory: a place near the South Pole built to study the night sky

microbe: a small living organism, like bacteria. Get out the microscope if you want to see one!

microscope: a device that makes small things looks bigger

North Pole: the location that is exactly the northernmost point on Earth

permafrost: soil that has been frozen for two years or more

tundra: a large, frozen plain where it's hard for things to grow

tussock: a clump of grass

Polar Science Resources

Tiny Ice: Bits from Antarctica, an English-language video series about life and science in Antarctica:
  https://www.polartrec.com/resources/informal-education-product/tiny-ice-bits-antarctica-video-series

Hielo Pequeño: Pedazos de la Antártida, a Spanish-language video series about life and science in Antarctica:
  https://www.polartrec.com/resources/informal-education-product/hielo-pequeno-pedazos-de-la-antartida-serie-de-videos

"10 Facts about the Arctic!" National Geographic Kids
  https://www.natgeokids.com/uk/discover/geography/general-geography/ten-facts-about-the-arctic

"What Is Permafrost?" NASA
  https://climatekids.nasa.gov/permafrost

Acknowledgments

PolarTREC

IceCube Collaboration and the Wisconsin IceCube Particle Astrophysics Center at the University of Wisconsin, Madison

ARCUS

Jocelyn Argueta is a writer and science communicator from Los Angeles, California. She is passionate about using the performing arts and books to inspire kids (and those young at heart) to be creative and curious about the world around them. Her favorite way to write is with a little help from the penguin lab assistant that sits on her desk.

Dr. Dieuwertje "DJ" Kast, EdD, is the director of STEM Education for the USC Joint Educational Project, based in Los Angeles, California. She holds a doctorate in education from USC, where she focused on teacher education in multicultural societies in STEM. Her mission is to level the playing field for underserved students in STEM.

Janet Pagliuca is a Venezuelan children's book illustrator and designer who found her passion for books at a young age. She graduated from Savannah College of Art and Design with a BFA in Communication Arts. Her inspiration comes from her personal experience growing up in two different cultures, which taught her about self-identity. She hopes to keep creating books that depict the importance of cultural awareness and diversity in children's storybooks.

Explore the Complete

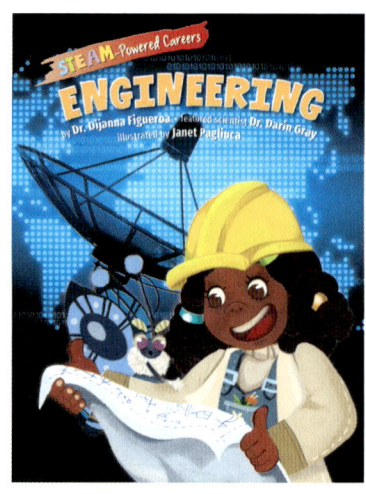

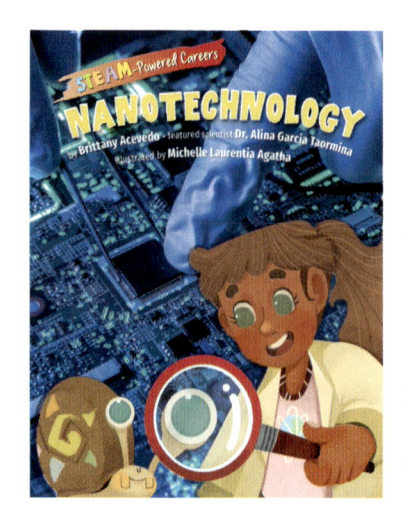

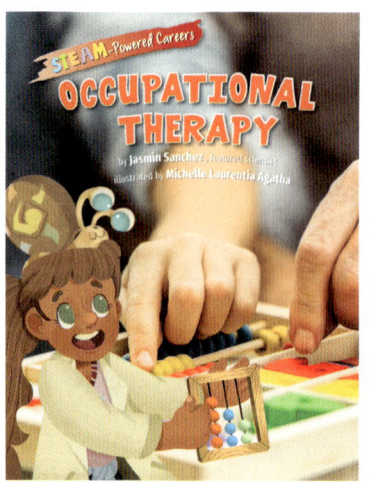

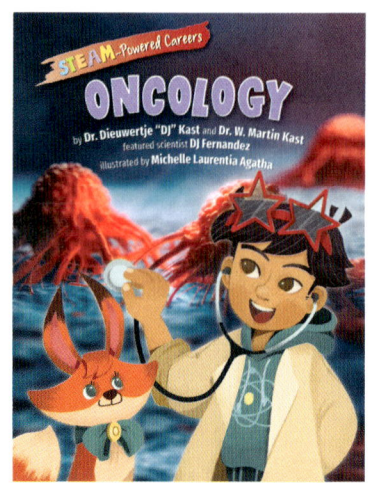

STEAM-Powered Careers Series!

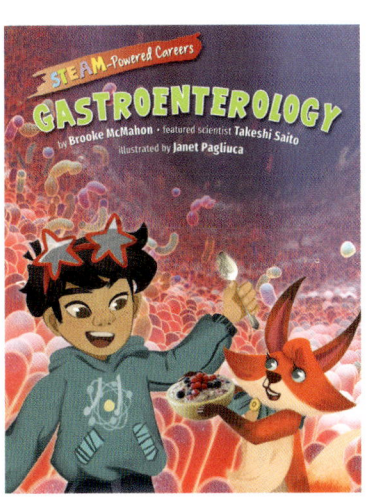

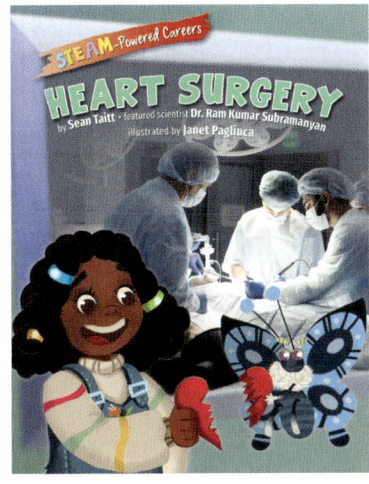

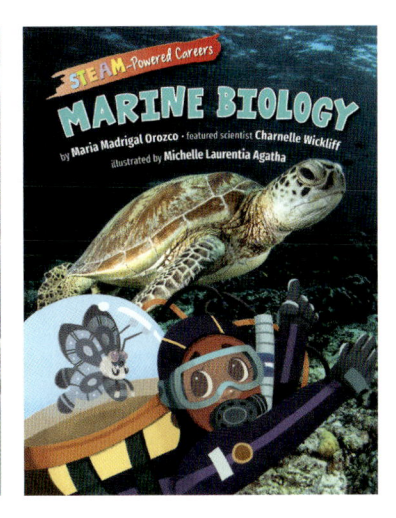

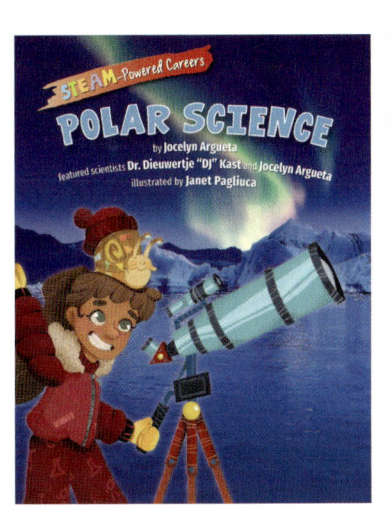

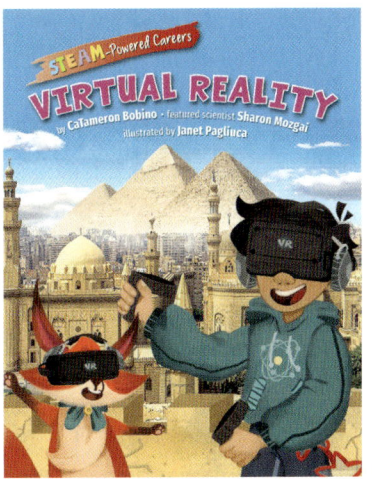

STEAM-Powered Careers: GASTROENTEROLOGY
by Brooke McMahon • featured scientist Takeshi Saito
illustrated by Janet Pagliuca

STEAM-Powered Careers: HEART SURGERY
by Sean Taitt • featured scientist Dr. Ram Kumar Subramanyan
illustrated by Janet Pagliuca

STEAM-Powered Careers: MARINE BIOLOGY
by Maria Madrigal Orozco • featured scientist Charnelle Wickliff
illustrated by Michelle Laurentia Agatha

STEAM-Powered Careers: POLAR SCIENCE
by Jocelyn Argueta
featured scientists Dr. Dieuwertje "DJ" Kast and Jocelyn Argueta
illustrated by Janet Pagliuca

STEAM-Powered Careers: VIRTUAL REALITY
by CaTameron Bobino • featured scientist Sharon Mozgai
illustrated by Janet Pagliuca

Photo credits

Cover JavierOlivares, CC BY-SA 4.0, via Wikimedia Commons **6** Blank Equirectangular Physical Map of the World, Mapswire.com, CC BY 4.0 **8–9** photos taken by Dr. Dieuwertje "DJ" Kast (PolarTREC 2016), courtesy of Arctic Research Consortium of the U.S.; Dave Stanley, CC BY 3.0, via Wikimedia Commons; Boris Radosavljevic, CC BY 2.0, via Wikimedia Commons **10** photos taken by Dr. Dieuwertje "DJ" Kast (PolarTREC 2016), courtesy of Arctic Research Consortium of the U.S. **11** Boris Radosavljevic, CC BY 2.0, via Wikimedia Commons **12** Boris Radosavljevic, CC BY 2.0, via Wikimedia Commons; S J Giovannoni, E F DeLong, T M Schmidt –N R Pace Appl. Environ. Microbiol. 1990, 56(8):2572 **14–15** photos taken by Jocelyn Argueta (PolarTREC 2019), courtesy of Arctic Research Consortium of the U.S.; Daniel Leussler, CC BY-SA 3.0, via Wikimedia Commons **16–17** photos taken by Jocelyn Argueta (PolarTREC 2019), courtesy of Arctic Research Consortium of the U.S. **18–19** Yuya Makino, IceCube/NSF **20** Blank Equirectangular Physical Map of the World, Mapswire.com, CC BY 4.0 **22–23** J. Richard Gott, David M. Goldberg, Robert J. Vanderbei; JavierOlivares, CC BY-SA 4.0, via Wikimedia Commons; stars by Chris Markhing from the Noun Project; Science by Made from the Noun Project; Telescope by Vectors Point from the Noun Project; Bacteria by farra nugraha from the Noun Project; Biology by Phatchara Bunkhachary from the Noun Project; Physics by Ilsur Aptukov from the Noun Project; Microbe by Path Lord from the Noun Project; Telescope by Jonas Nullens from the Noun Project; Bacteria by farra nugraha from the Noun Project **24** photos taken by Dr. Dieuwertje "DJ" Kast (PolarTREC 2016), courtesy of Arctic Research Consortium of the U.S., photos taken by Jocelyn Argueta (PolarTREC 2019), courtesy of Arctic Research Consortium of the U.S. **26–27** photos taken by Dr. Dieuwertje "DJ" Kast (PolarTREC 2016), courtesy of Arctic Research Consortium of the U.S.; © Evgeny Malkov | Dreamstime.com; Post of USSR, public domain, via Wikimedia Commons **28–29** photos taken by Jocelyn Argueta (PolarTREC 2019), courtesy of Arctic Research Consortium of the U.S.; Mike Lucibella, National Science Foundation; Amble, CC BY-SA 3.0, via Wikimedia Commons; photos taken by Jocelyn Argueta (PolarTREC 2019), courtesy of Arctic Research Consortium of the U.S.; B1mbo, CC BY-SA 4.0, via Wikimedia Commons; Manuel R.P., public domain, via Wikimedia Commons; Upload by Nickpo, public domain, via Wikimedia Commons **30–31** "Aufeis," Danielle Brigida, USFWS, CC BY 2.0, via Flickr **32–33** Noel Bauza/Pixabay **34–35** Chmee2/Valtameri, CC BY 3.0, via Wikimedia Commons **36–37** P J Hansen, CC BY-SA 2.0, via Wikimedia Commons; photo courtesy of Jocelyn Argueta; photo courtesy of Dr. Dieuwertje "DJ" Kast; photo courtesy of Janet Pagliuca **40** "Antarctica: South Pole Telescope," Eli Duke, CC BY-SA 2.0, via Flickr